It was Job Day in Miss Hill's classroom. Miss Hill said, "I'm glad you could all come. I know we will enjoy finding out about your careers."

1

Dawn's mom said, "I'm a cook at a diner. I unfreeze food in the morning. Later I precook some of the food. Then I preheat the grill and fix meals."

Martin's father said, "I'm a bookshop owner. Every morning at eight I unlock my shop. I put the sale books on a discount shelf. Sometimes I try to give helpful tips to shoppers."

Van's mother said, "I'm a gardener. I work at the biggest park in town. I take care of plants. The park is a peaceful place to visit."

Sandy's older sister unzipped her bag. "I teach art to preschool classes," she said. "I'm also a painter. I premix paints to make colorful paintings like this."

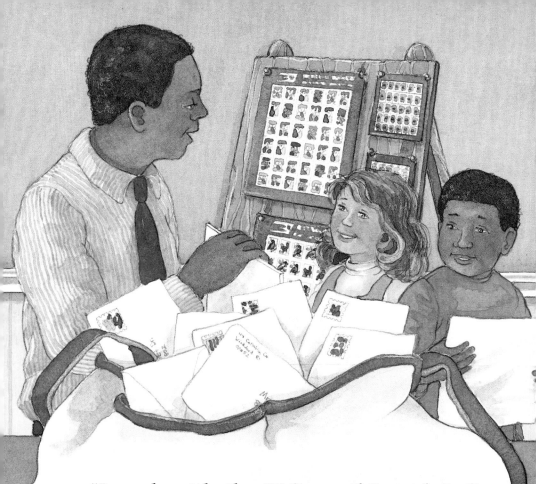

"I work with the U.S. mail," said Jed's dad. "I help unload the mail trucks. I presort letters and other mail. Then other workers take the mail to people's homes."

"I'm a baker," said Lin's granddad.
"I make bowlfuls of cookie mix. I drop
spoonfuls of it onto cookie sheets. Out
of the stove come the sweetest cookies
in town!"

Miss Hill said, "Those sound like
great careers. We are thankful that
you all came." All the students clapped.